Velocity is a measure of an object's speed in a particular direction.

Acceleration is the rate at which velocity changes.

READING FOCUS SKILL
MAIN IDEA AND DETAILS

The main idea is what the text is mostly about. Details are pieces of information about the main idea.

Think about the main idea of each page, and look for details about how motion changes.

Forces and Motion

How can you tell if something is moving? You can compare the object with something that is not moving. This is called a *frame of reference*. For example, you can tell that a car is moving if you see it pass a tree. You have to look at both the car and the tree. You look at objects inside a frame.

How can you define motion? An object is in motion when it changes its position. **Position** is an object's location. We use words like *east*, *behind*, *left*, *right*, and *above* to describe position.

The position of these girls changes from the top of the slide to the bottom.

A push is a force that keeps a swing in motion.

Forces can make objects change position. All forces are pushes or pulls.

Gravity is a force that pulls objects toward Earth. The pull of Earth's gravity keeps your chair on the ground.

Friction is a force that opposes motion. It slows down or stops motion by acting on the surface of an object. The smoother the surface is between two objects, the less the friction.

Forces can work together to affect the way an object moves. An airplane in flight is pulled to the ground by *gravity*. But a force called *lift* pushes it up. *Drag* from air slows the plane. But *thrust* from the engines pushes the plane forward.

 What are some ways forces can affect the motion of an object?

Speed and Velocity

Some things, like airplanes, move very fast. Some things move so slowly you can hardly tell they are moving.

How can you describe how fast something is moving? You can see how long it takes for it to move a distance. This measurement is called speed. **Speed** is the distance an object travels in a certain amount of time. "Thirty miles per hour" is an example of speed. An object's speed tells you how fast or slowly the object is moving.

High-speed trains can travel hundreds of miles an hour. ▼

Motion

Lesson 1
What Factors Affect Motion?...................... 2
Lesson 2
What Are the Laws of Motion?.................... 10
Lesson 3
How Do Waves Move Through Different Materials?... 20

Orlando   Austin   New York   San Diego   Toronto   London

Visit *The Learning Site!*
www.harcourtschool.com

Lesson 1

What Factors Affect Motion?

VOCABULARY
position
speed
velocity
acceleration

Position is the location of an object. We use words like *north*, *south*, *east*, *west*, *above*, *below*, and *behind* to tell position.

Speed is the distance an object travels in a certain amount of time. Speed tells you how fast or slowly something is moving.

The speed of an object tells you how long it will take to move a certain distance. Say a car is going 50 miles per hour. You know it will travel 100 miles in two hours. But you don't know the car's direction.

Sometimes you also need to know the direction an object is going. **Velocity** measures an object's speed in a certain direction. One car's velocity may be 50 miles per hour east. Another car's velocity may be 50 miles per hour west. They both have the same speed. But they have different velocities.

 What is the difference between speed and velocity?

▼ **The velocity of the cheetah describes its speed and direction.**

Acceleration

What happens when a moving object changes direction? Its velocity changes. Its velocity also changes if the object slows down or speeds up. We call this change in velocity acceleration. **Acceleration** is the rate at which velocity changes.

Some car ads talk about acceleration. The ad might say "0 to 60 in 5 seconds." This means that the car can change its velocity from 0 miles per hour to 60 miles per hour in 5 seconds.

 What does an object's acceleration describe?

▲ Acceleration describes both the way this plane speeds up to take off and how it slows down to land.

Momentum

Momentum describes how easy or hard it is to stop a moving object. Momentum depends on how much mass the object has. It also depends on the object's velocity. At the same velocity, a heavy truck will have more momentum than a small car. If two cars have the same mass, the car with higher velocity has more momentum. The more momentum a moving object has, the more force and time it will take to make it stop.

 Why do objects with more mass have greater momentum?

This car has a certain mass, ▶ but its momentum can change if its velocity changes.

Review

Complete this main idea statement.

1. The motion of an object changes when a _____ such as friction acts on it.

Complete these detail statements.

2. The distance an object travels in a certain amount of time is its _____.

3. The rate at which the velocity of an object changes is its _____.

4. Two factors that affect an object's momentum are its mass and _____.

Lesson 2

VOCABULARY
inertia

What Are the Laws of Motion?

Objects tend to resist a change in motion. This is **inertia**. Inertia keeps you moving forward if a car stops suddenly. Safety belts apply the force needed to stop you.

READING FOCUS SKILL
CAUSE AND EFFECT

A cause is what makes something happen. An effect is what happens.

As you read, look for cause and effect relationships in the way the laws of motion work.

Newton's First Law of Motion

Isaac Newton described how things move. He developed three laws of motion.

Newton's first law of motion describes inertia. **Inertia** is the tendency of objects to resist a change in motion. This means that moving objects tend to keep moving. Objects that are not moving tend to stay still.

 Why does the golfer have to apply a force to make the ball move?

◀ The golf ball stays still because of inertia.

Newton's Second Law of Motion

Newton's first law tells us that if you kick a ball, the ball will move. Kicking it overcomes the force of inertia that keeps the ball at rest.

But what happens when you change the amount of force you use? If you kick a ball gently, it will move slowly. If you kick it hard, the ball will move faster.

Newton's second law says that an object's acceleration depends on the force you apply to it.

The ribbon requires only a little bit of force to accelerate because it has a small mass.

Acceleration is the rate at which velocity changes. The more force you apply, the greater the acceleration. So the harder you kick a ball, the faster the ball goes.

A car driver can tap the gas pedal gently. This makes the car move slowly. Another driver can "floor" it. This driver applies more force. So the car accelerates faster.

The bicycle's acceleration depends on how much force the rider applies to the pedals.

Newton's second law also includes mass. It says that an object's acceleration depends on its mass.

Suppose you and your friend must move two boxes of books that have the same mass. You have to move them the same distance. You push against one box and find that it takes a lot of force to move. Your friend takes half the books out of the other box. Your friend can move this box with much less force. This is because it has less mass.

You can also see how Newton's second law works at the grocery store. An empty cart is much easier to push than a full cart. The person in the picture below must use a lot of force to move all of these carts. He would need much less force to move one cart. This is because it is easier to accelerate objects with less mass. One cart has much less mass than all the carts together.

 What effect does mass have on an object's acceleration?

The long row of empty shopping carts requires more force to move than one empty cart takes.

Newton's Third Law of Motion

Newton saw that forces occur in pairs. When one force acts, another force acts against it. It acts in the opposite direction. One way of saying Newton's third law is to say that for every action force, there is an equal and opposite reaction force.

When you push on a heavy box, you apply an action force. The box pushes back with a reaction force. The box will move only if you push with more force than the box pushes on you.

 How do you know a heavy object exerts force?

action force | reaction force

As the boy pushes off the boat, the boat pushes back on him in the opposite direction.

17

Motion in Space

Have you seen pictures of astronauts in space? You might see them floating in the space shuttle. You might also see objects floating or spinning around them.

It looks as though gravity does not act in space. But gravity acts in space as much as it does on Earth. The difference is that, in space, gravity is balanced by inertia. Inertia moves the shuttle forward. And the shuttle moves fast enough so gravity does not make it fall to Earth.

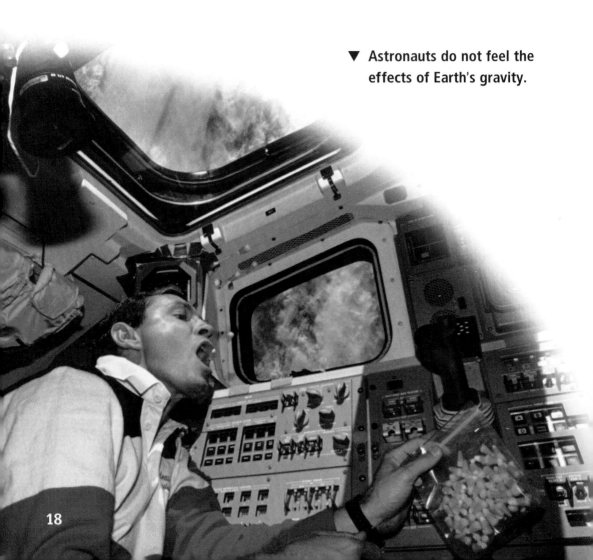

▼ Astronauts do not feel the effects of Earth's gravity.

The same balance of forces keeps the moon in orbit around Earth and Earth orbiting the sun.

◀ Drag acts on this space capsule as it reenters Earth's atmosphere.

The satellite orbits Earth. ▶
It's pulled toward Earth by gravity.

Review

Complete these cause and effect statements.

1. The property of _____ causes a moving object to keep moving unless other forces act on it to slow or stop it.

2. The effect of Newton's _____ law of motion is seen when a lighter object accelerates faster than a heavier object when the same force is applied.

3. When you push on a heavy object, the object pushes back with an equal and _____ force.

Lesson 3

How Do Waves Move Through Different Materials?

VOCABULARY
vibration
volume
pitch
frequency

A **vibration** is a back-and-forth movement of matter. The drum's thin covering vibrates when someone hits it.

The loudness of a sound is called **volume**. A jackhammer makes a sound loud enough to hurt your ears.

The **pitch** of a sound is how high or how low it is. The pitch of the sound each glass makes depends on how much water is in it.

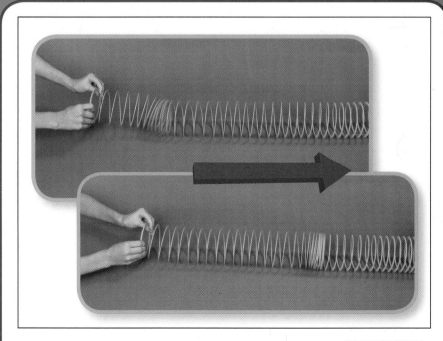

The number of vibrations per second is the **frequency** of a sound.

READING FOCUS SKILL
MAIN IDEA AND DETAILS

The main idea is what the text is mostly about.
Details are pieces of information about the main idea.

As you read, look for information about what sound is and details about how sound travels.

Sound Energy

Energy has many forms. One form is sound energy. Sound energy travels through the air.

Sound is made when something vibrates. A **vibration** is a back-and-forth movement of matter. The vibrations make the air vibrate. This vibration is what helps you to hear.

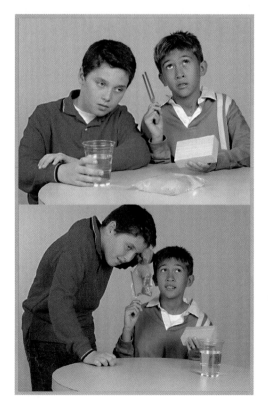

◀ These boys hear sound from the vibration of the tuning fork.

Musical instruments make sounds in different ways. A drum makes sound when you hit it. A guitar makes sound when you pluck the strings. A clarinet makes sound when you blow into it. This makes a wood reed vibrate.

The loudness of a sound is called the **volume**. The volume of a sound is measured in *decibels (dB)*. A high-decibel sound is loud and has a lot of energy.

 What are three ways you can make sounds with a musical instrument?

How Loud Are Some Sounds?	
Sound	**Decibel Level**
Whisper	20 dB
Quiet radio	40 dB
Conversation	60 dB
Dishwasher	80 dB
Jackhammer	100 dB
Thunderclap	120 dB

Sound Waves

Sound moves through the air as waves. When you hit a drum, the cover vibrates and so does the air directly above it. This squeezes, or compresses, molecules of air together. The compressed air squeezes the air next to it. In this way, the vibration moves through the air.

Some sounds are higher than others. A trumpet makes a higher sound than a tuba. A sound's **pitch** is how high or how low it is.

The number of vibrations per second is the **frequency** of a sound. If you pluck a long string and a short string, the short string will vibrate faster. This means that the short string has a higher frequency than the long string has.

▼ The springs model how sound travels in compressed waves.

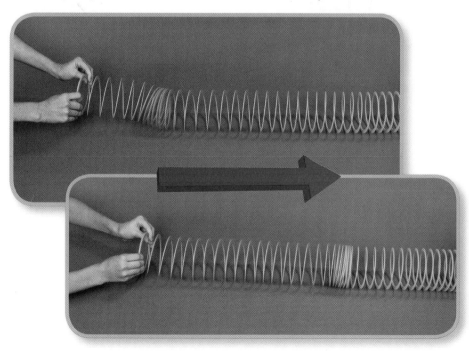

Frequency and pitch are related. A sound with a high frequency has a high pitch. A sound with a low frequency has a low pitch.

Sound waves move out in all directions from an object. When a sound wave hits something, some of the energy is absorbed. Soft surfaces absorb more energy than hard surfaces. A sound that hits a hard surface bounces back. A sound that bounces off a hard surface is called an *echo*. You can often hear echoes in caves and canyons.

 How is pitch related to frequency?

▼ Sound bounces off the hard cliff and produces an echo. The echo is not as loud as the original sound.

Sound Movement

Think about doing "the wave" in a stadium. Fans stand up and sit down at their seats. Fans don't move around the stadium. But the wave's energy can travel a long distance. Sound waves move like that. They travel through the air because molecules of air vibrate in place. Sound energy can travel a long distance. But the air stays in one place.

Any kind of matter can vibrate and carry sound. The speed of sound depends on what it travels through. Sound travels fastest in solids. It travels slowest in gases. Sound also moves faster in warmer temperatures than in colder temperatures.

 Sound waves from the singer travel in all directions to the audience.

When the performer sings, she produces vibrations that compress the air.

The sound moves through the air as waves.

The sound waves reach the people in the audience, and they hear the singer.

Animals and Sound

Many animals can hear sounds that humans cannot hear. Dogs can hear high-pitched sounds that people cannot hear.

Bats have excellent hearing. When a bat flies, it produces many sounds. These sounds bounce off objects. The bat hears the echoes. The echoes give the bat information. This allows the bat to fly at night.

The bat uses echoes to find prey.

Review

Complete these main idea statement.

1. _____ travels through matter as a compression wave.

Complete these detail statements.

2. A sound with a _____ pitch has a high frequency.

3. When sound waves strike a soft surface, like a carpet, most of the sound energy is _____.

4. Sound travels faster in _____ , and slowest in gases.

GLOSSARY

acceleration (ak•sel•er•AY•shuhn) the rate at which velocity changes.

frequency (FREE•kwuhn•see) the number of vibrations per second.

inertia (in•ER•shuh) the property of matter that tends to keep it at rest or moving.

pitch (PICH) how high or low a sound is.

position (puh•ZISH•uhn) the location of an object in space.

speed (SPEED) the distance an object travels in a certain amount of time.

velocity (vuh•LAHS•uh•tee) a measure of an object's speed in a particular direction.

vibration (vy•BRAY•shuhn) a back-and-forth movement of matter.

volume (VAHL•yoom) the loudness of a sound.